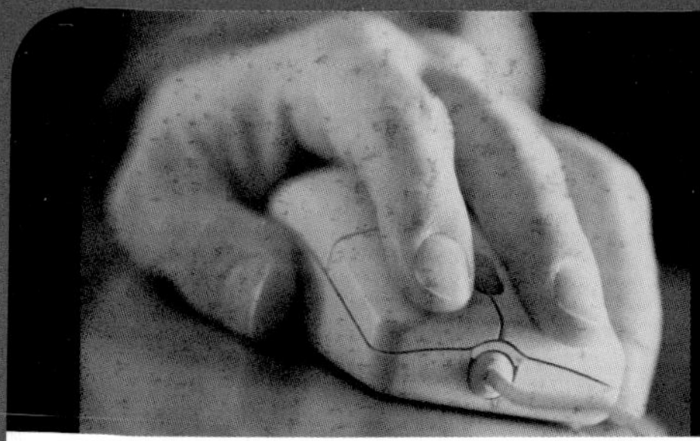

Log on.

Tune in.

Succeed.

Your steps to success.

STEP 1: Register

All you need to get started is a valid email address and the access code below. To register, simply:

1. Go to www.ConceptualSciencePlace.com
2. Click the appropriate book cover.
   Cover must match the textbook edition being used for your class.
3. Click "**Register**" under "**First-Time User?**"
4. Leave "**No, I Am a New User**" selected.
5. Using a coin, scratch off the silver coating below to reveal your access code.
   Do not use a knife or other sharp object, which can damage the code.
6. Enter your access code in lowercase or uppercase, without the dashes.
7. Follow the on-screen instructions to complete registration.
   During registration, you will establish a personal login name and password to use for logging into the website. You will also be sent a registration confirmation email that contains your login name and password.

Your Access Code is:

Note: If there is no silver foil covering the access code, it may already have been redeemed, and therefore may no longer be valid. In that case, you can purchase access online using a major credit card. To do so, go to www.ConceptualSciencePlace.com, click the cover of your textbook, click "**Buy Now,**" and follow the on-screen instructions.

STEP 2: Log in

1. Go to www.ConceptualSciencePlace.com and click the appropriate book cover.
2. Under **"Established User?"** enter the login name and password that you created during registration. *If unsure of this information, refer to your registration confirmation email.*
3. Click **"Log In."**

STEP 3: (Optional) Join a class

Instructors have the option of creating an online class for you to use with this website. If your instructor decides to do this, you'll need to complete the following steps using the Class ID your instructor provides you. By "joining a class," you enable your instructor to view the scored results of your work on the website in his or her online gradebook.

To join a class:

1. Log into the website. For instructions, see "STEP 2: Log in."
2. Click **"Join a Class"** near the top right.
3. Enter your instructor's **"Class ID"** and then click **"Next."**
4. At the Confirm Class page you will see your instructor's name and class information. If this information is correct, click **"Next."**
5. Click **"Enter Class Now"** from the Class Confirmation page.

- *To confirm your enrollment in the class, check for your instructor and class name at the top right of the page. You will be sent a class enrollment confirmation email.*
- *As you complete activities on the website from now through the class end date, your results will post to your instructor's gradebook, in addition to appearing in your personal view of the Results Reporter.*

To log into the class later, follow the instructions under "STEP 2: Log in."

Got technical questions?

Visit http://247.aw-bc.com. Email technical support is available 24/7.

SITE REQUIREMENTS

For the latest updates on Site Requirements, go to www.ConceptualSciencePlace.com, choose your text cover, and click Site Reqs.

WINDOWS
366 MHz CPU
Windows 98/2000/XP
64 MB RAM
800 x 600 screen resolution, thousands of colors
Browser: Internet Explorer 5.0/5.5/6.0; Netscape 6.2.3 and 7.0
Plug-Ins: Shockwave Player 8, Flash Player 7, QuickTime 5, Java
Internet Connection: 56k modem minimum

MACINTOSH
266 MHz PowerPC
OS 9.2/10.2.4/10.3.2
64 MB RAM
800 x 600 screen resolution, thousands of colors
Browsers: Internet Explorer 5.1/5.2; Netscape 6.2.3; Safari 1.2*
Plug-Ins: Shockwave Player 8, Flash Player 7, QuickTime 5, Java
Internet Connection: 56k modem minimum

* The gradebook and class manager features of this website have been partially tested on Mac OS X 10.3.2 with Safari 1.2. A future update will fully support this configuration.

Join a Class

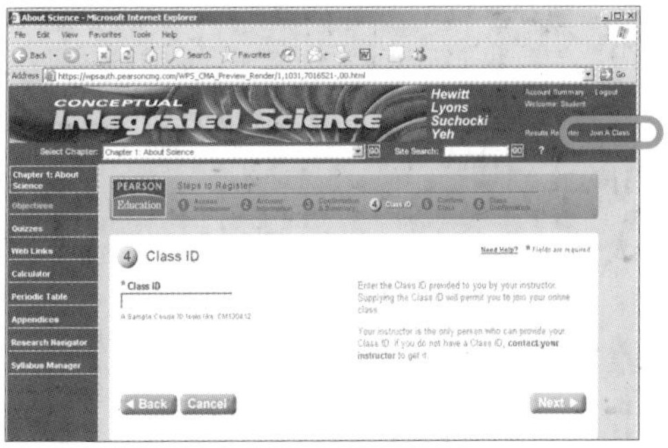

Register and log in

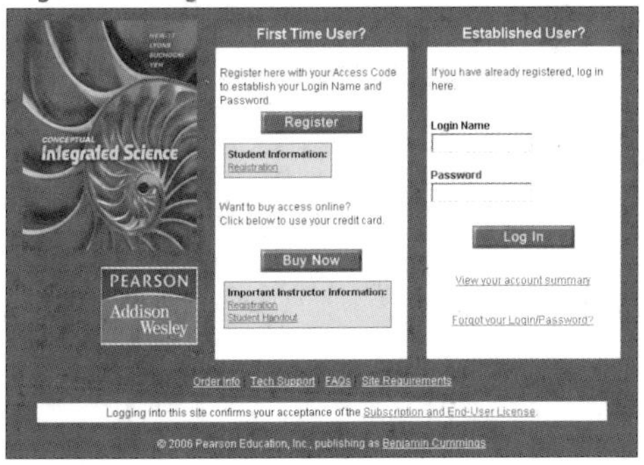

Important: Please read the Subscription and End-User License Agreement, accessible from the book website's login page, before using *The Conceptual Science Place* website. By using the website, you indicate that you have read, understood, and accepted the terms of this agreement.

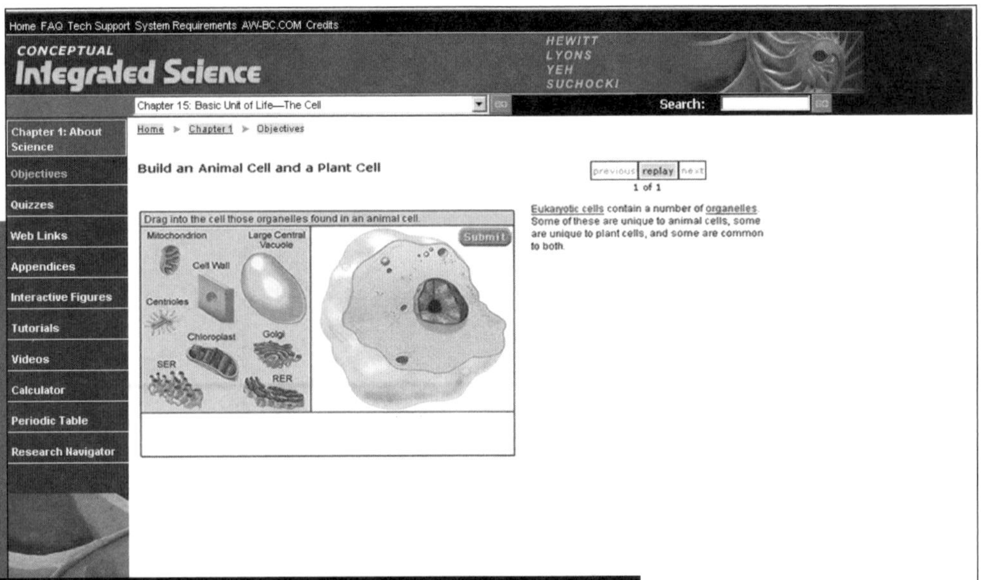

This grid is your key to The Conceptual Science Place website, designed to make learning easy and fun!

This media grid provides an at-a-glance, chapter-by-chapter overview of how the text and media work together to illustrate key concepts. The grid outlines the wealth of interactive figures, interactive tutorials, and videos available for use with *Conceptual Integrated Science.* In addition, for every chapter of the text, the website includes self-assessment quizzes, flashcards, links, and more.